OIL!

GETTING IT
SHIPPING IT
SELLING IT

OIL!

GETTING IT

SHIPPING IT

SELLING IT

ELAINE SCOTT

FREDERICK WARNE
NEW YORK LONDON

The author wishes to thank Elizabeth Jogeese of the International Association of Drilling Contractors and Princetta Johnson of the Exxon Corporation for their gracious assistance in the photograph research.

Frederick Warne & Co., Inc.
New York, New York

Printed in the U.S.A. by The Murray Printing Company
Book design by Mina Greenstein

1 2 3 4 5 88 87 86 85 84

Library of Congress Cataloging in Publication Data
Scott, Elaine.
Oil! : getting it, shipping it, selling it.
Summary: Describes the processes which bring oil to our home—searching, drilling, pumping, shipping, and refining. Also discusses the problems of our dependency on oil and alternate energy sources.
1. Petroleum industry and trade—Juvenile literature.
[1. Petroleum industry and trade]   I. Title.
HD9560.5.S373   1984   338.2′7282   83-19653
ISBN 0-7232-6260-8

FOR MY SISTER KATHY
whose friendship and love mean more to me
than a field of pumping wells

Spudding In 1

Hundreds of thousands of years ago, when the earth was young and dinosaurs were alive, long before anyone wrote anything down—long before there *was* anyone—oil had been formed and was lying in pools deep inside the earth. In fact, some of the oil was lying on top of the earth's crust, in thick and sticky puddles. Centuries passed, and eventually men and women began to give names to the objects in the world around them. They named this oozing, smelly, sticky substance *bitumen* (today we call it *asphalt*), and they noticed that it had many uses. For example, our ancestors learned that bitumen burns, so it could be used to cook food or to make a fine torch. When it is spread out, it hardens, so it could be used as a pavement. The streets of ancient Babylon were paved with bitumen. It was used also as a mortar, or glue, to hold pieces of tile and stone together for buildings and the decorations that went on them. And because it is waterproof, bitumen was used to caulk the boats of ancient warriors—and the boat of one special baby.

At first, people passed on their history by telling stories around a campfire. Eventually, however, they began to write these stories down, and the world had a written history. Some of the world's oldest stories are recorded in the Bible. There, the book of Genesis tells about the Israelites using bitumen to hold the Tower of Babel together. In Exodus, we read the story of Moses and how his mother wove a boat of bullrushes (a type of reed), using bitumen to waterproof it, then placed her son in it and floated him in the river to escape the pharaoh's army.

Moses' story took place around 1290 B.C. during the reign of Pharaoh Ramses II of Egypt, so we know that oil has been useful to mankind for at least three thousand years. But no one drilled for oil. Early man dug water wells by hand, but he did not dig oil wells. Oil was just there, sitting on top of the earth. People gathered it by skimming it off the top of the puddles or by dipping blankets into the puddles, then wringing the oil out.

The Chinese were probably the first people to drill, rather than dig, for both water and oil. As long ago as 600 B.C., the Chinese were using bits of bronze and pieces of bamboo to drill into the earth in search of the treasures, such as oil and water, that were hidden inside. By 1500 A.D., the Chinese could drill to a depth of 2,000 feet. Just 125 years ago, in 1859, the first oil well in the United States came in at 69 feet. Today, half a world away and nearly 2,500 years after the Chinese first drilled, men and women in Oklahoma are drilling into the earth, too, searching for the treasure of oil and the gas that often floats on top of it. But their drills are made of steel, not bamboo, and they are powered by machinery, not muscles. In Oklahoma, they are drilling to a depth 33,000 feet—that's more than *six miles*—sinking a well from which they hope they will be able to pump natural gas and oil—gifts from nature that make life more comfortable and safer for everyone on our planet.

The ancient Chinese people probably did not worry about money and how much it cost to drill their wells. Bamboo was plentiful, and they used muscles instead of machines to power their drills. Today, it is very expensive to drill for oil or gas.

The six-mile-deep Robinson well in Oklahoma, the world's deepest gas well, will cost between 23 and 25 *million* dollars before it is finished, and there is no guarantee that there will be anything but a dry hole when the drilling is completed. However, the company drilling the Robinson well must be willing to take that chance. The only way the world will get the oil and gas it needs is to drill for it; the oil puddles are gone forever.

Not all wells cost as much as the Robinson well, but any well costs at least hundreds of thousands of dollars. Naturally, since it costs that much, the oil companies that are drilling the wells want to be fairly certain that they have a good chance of finding oil when they get to the bottom. So they employ scientists called *geologists* and *geophysicists* to study the earth and decide if a particular location is likely to have oil trapped below.

At this point it might be helpful to understand how oil was formed and got trapped in the earth in the first place. The following explanation is a *scientific theory*, which means that most scientists believe it is true based on the evidence, but no one can prove absolutely that it *is* true. Oil is called a *fossil fuel* because scientists believe it had its beginnings in the plants and animals that lived on the earth over 500 million years ago. Most of these ancient animals and plants lived in the sea, which covered much of the earth at that time. When these creatures died, their remains sank to the bottom of the sea and decayed, and as they decayed, their remains produced fats and oily substances, which settled into the layers of sand on the ocean's floor. Ancient rivers also carried bits of earth and decayed animal and plant life from the earth to the sea, and these fragments drifted to the ocean's floor, too, laying down new layers of sand and mud sediment. Gradually, the

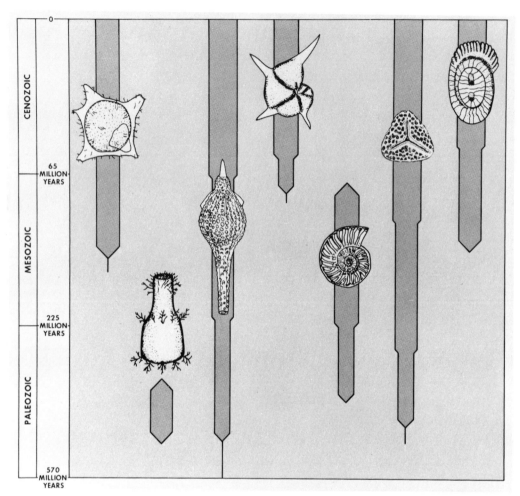

Scientists can determine the age of sedimentary rock layers by studying the kinds of fossils they find in those layers. These are a few typical fossils and their ages.

weight of each new layer of sediment compressed the older, deeper layers into porous, spongelike rocks, such as sandstone or limestone. These kinds of rocks are called *sedimentary rocks* because they were formed by the pressure of all of those layers of ancient sand and mud sediment.

The pressure from each new layer of sediment, along with the heat from inside the earth, affected the fats and oily substances from the decayed plant and animal life, too. Slowly, these substances were changed into the liquid we call crude oil. As the oil was formed, it settled into the holes in the sedimentary rocks. Through millions of years, the earth has shifted and moved, heated and cooled. As it did, the sedimentary rock layers shifted and moved, too, trapping the oil in pockets deep inside the earth. The oil that didn't get trapped in the earth was squeezed to the top and formed the oil puddles we have already discussed.

A slight hill, or bulge in the earth's surface, could indicate that oil is trapped below in layers of rock that, due to shifts in the earth's surface, have arched upward. This kind of trap is called an *anticline.* Sometimes pressure makes the earth fracture and the underground layers of rock shift, causing a *fault,* which also can trap the oil. At other times underground pillars of salt called *salt domes* form, trapping the oil and keeping it below. So now we know how the oil came to be trapped. But *where* is it trapped? the oil companies ask. They look to geologists and geophysicists to provide the answer.

The geologists, whose specialty is studying the earth, begin by looking at photographs of the earth's surface that have been taken from an airplane. Sometimes they can tell if an area might have oil just by looking at the way the land bumps and curves. If the geologists think they see something in the photographs that indicates an anticline, a fault, or a salt dome could be there, they will drill down and bring up a plug of earth that contains a sample of the rock that lies below. The geologist studies this plug, which is called a *core sample,* to see if

This seismograph records the sounds the earth makes as it shakes after small man-made earthquakes are set off.

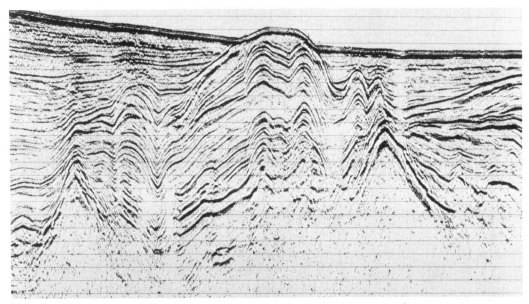

A seismic record gives a picture of the underground rock layers. The anticline shown here could indicate that oil is trapped underneath the arched rock.

it contains sedimentary rock. If it does, the geophysicist could be called in next.

Geophysicists are scientists who study how the earth moves. They set off man-made earthquakes, then instruments called *geophones* listen to the sound the earth makes as it shakes. The sound is converted to electricity and is sent to an instrument called a *seismograph*, which records the sounds as a *seismic record*. A seismic record is a kind of picture of the underground rock layers. By looking at these seismic records, geophysicists can tell if there are anticlines, faults, or salt domes buried beneath the earth's crust at the spot where the oil company wants to drill.

However, even with core samples and a seismic record, no one can know for certain if oil is there until the bit of the drill punctures the top of the trap. The drilling process can last for weeks or, in the case of the Robinson well in Oklahoma, even years. Although the knowledge and experience of the geologist and geophysicist certainly improve them, the odds for success in drilling an oil well are not very good. Out of every ten wells that are drilled, only one will contain oil. And only one well out of fifty will contain enough oil to make it worth pumping to the surface. So you can see that drilling an oil well is a chancy proposition.

There are two major methods of drilling for oil—*cable drilling* and *rotary drilling.* Cable drilling, sometimes called *percussion drilling*, is the oldest method—the Chinese used a form of it. In this method, the drill literally beats, or hammers, the hole into the earth, so you can see how it got the name percussion. The drill bit, the actual cutting tool that bites into the earth, is lifted and dropped over and over again, pounding the well hole into the earth. In the early days of the oil industry, the cables that raised and dropped the drill bit were actually ropes made out of hemp, a plant that has fibers in its stalk. However, as time went on and the wells grew deeper, the drills had to be lifted higher and higher just to get them out of the hole, so the rope cables had to be thicker. Eventually, in order to have the needed strength, they would have been too thick to be practical, so cables made of twisted strands of metal were introduced. Metal cables are stronger and thinner than rope cables and are used for cable drilling today.

Cable drilling is still used in various parts of the world, especially in those locations where the well will not be very deep. However, most of the new oil wells being drilled use rotary drilling techniques. As the name implies, in rotary drilling the drill turns, or rotates, boring the hole into the ground instead of pounding it in.

_____

Rotary drilling has two advantages over cable drilling. A rotary rig can drill deeper, and through softer rock formations, than a cable rig. Much of the oil that we find today lies in traps that are located deep within the earth, so there are far more rotary rigs in action today than cable rigs. Then too, a rotary rig drills faster than a cable rig. The cost of drilling an oil well is usually determined by the time it takes to drill the well— the faster you drill, the cheaper the well—so in most cases, it's less expensive to work with a rotary rig. We will look at a rotary rig in action.

The first well that is drilled at a particular site is called an *exploratory well*. If oil is found, more wells will probably be drilled there on the land the oil company has leased from its

A modern drill bit. The holes at the top let the drilling fluids circulate through the bit to cool it down.

owner. Sometimes individual people own the land, and sometimes the land is owned by the government of a nation. No matter who owns it, the land must be leased by the oil company before the company can start any of its work, from taking aerial photographs to sinking the well.

When the oil company is ready to begin drilling, the land around the site for the well is cleared of any trees and bushes that would interfere with the drilling process. If the well site is very far away, roads leading to it may have to be built, and the oil company takes care of this job. Trucks holding the derrick (the towering framework that fits over the mouth of the oil well) and other pieces of heavy equipment roll into place. Because water is essential to a drilling operation, either it will have to be piped in or a water well will have to be dug. Waste pits for the debris from the well are dug, and offices in trailers are set up.

The mighty derrick and the production platform go up next. In the old days, the derricks were made of wood and were constructed at the site of the well. Today, most oil companies use what is called a jackknife derrick, which can be folded and put on a truck—or trucks if the derrick is very large—for delivery to the well site. At a very remote well site, the derrick may be flown in by helicopter in pieces and then assembled at the well site. The derrick supports the weight of the *drill string*, which is lengths of drill pipe that have been joined together. Unlike the ancient Chinese, today's drillers make drill string not of bamboo but of steel. The *drill bit*, the actual cutting tool, is attached to the end of the drill string by a fitting called a *drill collar*. If the well is deep and many lengths of pipe have to be joined, the total weight of the

In America the earliest derricks were made of wood.

Above: A modern steel derrick rests on a production platform.
Right: A close-up look at a production platform in Texas.

drill string, drill collar, and bit can be as much as 500,000 pounds. The derrick must be able to support this weight and also withstand winds of one hundred miles an hour should they sweep across the land. An average derrick rises about 110 feet above the *production platform*, which is simply the surface on which the people who actually drill the oil well work.

With the derrick erected, the men are ready to "spud in"— oil-field talk for the moment they start the hole that will become, they hope, a producing well. In rotary drilling, the first step in spudding in is to insert a length of pipe, called a *conductor pipe*, into the earth and cement it into place. The conductor pipe serves two purposes. It keeps the loose dirt and rocks that are near the earth's surface out of the new oil well, and, if there is any ground water near the well, the conductor

Above: Drill pipe waiting to be joined to make drill string.
Right: The driller, shown here at his instrument panel, is responsible for all the other workers on the rig.

pipe keeps it pure and clean, too. The bit will bite into the earth through the center of this pipe. Once the work begins, the job of drilling for oil will continue around the clock, twenty-four hours a day, seven days a week, until the drill bit finds oil—or nothing—in the trap below.

People in the oil industry have a unique and colorful vocabulary. For example, the person in charge of the entire drilling operation at a drill site is called the *tool pusher*. In the early days of the oil industry, it was the tool pusher's job to keep the workers supplied with the proper tools, so it is easy to see how he came to be called a tool pusher.

Each day the tool pusher receives the reports of the *driller*, the person who is in charge of the men and women who actually do the work of drilling the oil well. The work is hard and dangerous and is divided up into three eight-hour shifts, or tours, a word that is pronounced (and occasionally spelled) "towers." The driller is in charge of all the other workers on his or her tour. Each day the three drillers, one for each tour, report the progress of the well to the tool pusher, including the information on how much "hole" the crew made during its tour. Everyone connected with the oil well wants the drilling to be completed as quickly as possible, and many drillers pride themselves on the fact that the men and women on their tour drilled deeper and faster than those on the tour before.

Above: "Making a trip" on an oil well in California.
Right: A new drill string is "run back in."

When the drilling process begins, the drill bit and drill collar are attached to one end of a piece of drill pipe. A piece of drill pipe is about thirty feet long, so when the driller reaches that depth, he has to add more pipe if he wants to keep on drilling. Two or more pieces of drill pipe joined together become the drill string. The person on a drilling crew who joins the lengths of pipe together is called a *derrickman*, though some derrickmen are women. When a drill string has two lengths of pipe on it, it is called a *double*. When another length of pipe is added, the drill string becomes a *thribble*, and four lengths of drill pipe joined together are called *fourbles*. If the

drill string needs to be pulled out of the hole for any reason, it is broken up into doubles, thribbles, or fourbles to be hoisted. A small rig with a short derrick can hoist a double, and only a very large rig with a very tall derrick can manage to hoist a fourble out of the hole. Nevertheless, no matter how deep the well, the entire drill string can be hoisted from it. When the well is very deep, many foibles are hoisted before the entire drill string is out of the hole. When the driller pulls the drill string out of the hole, he says he is "making a trip." In the case of a deep well, a trip can take as much as fifteen hours to complete!

Different kinds of drill bits, ranging in size from 5 to 9 inches across, are used to break up different kinds of earth. As they grind through the earth, they wear down and must be replaced. It is not unusual for one drill bit to cost around $8,000, and several will be used in the process of drilling the well. If the earth is soft, the bit has relatively large steel teeth. As the drill plunges down through harder rock layers, the bit will be changed to one with smaller teeth. Some bits, the ones that are used to drill through the hardest rock formations, have teeth made of industrial diamonds.

In addition to being raised and lowered so the bits can be changed, the drill string must also be able to rotate as the well is drilled. Two pieces of drilling equipment allow the rotating action to happen—the *swivel* and the *kelly*. The swivel is attached on one end to the *traveling block*, the largest pulley on the drilling rig. The other end of the swivel is attached to the kelly. The kelly is not round like ordinary pipe. It has at least four, sometimes more, sides to it, and it provides the *torque*, the twisting or rotating motion, that the drill string must have. The kelly fits into a slot in the *rotary table* on the rig's production platform, and then the drill string itself is attached to the kelly. When the driller signals for the drilling to begin, the turntable rotates, turning the kelly and the attached drill string, and the drill bit takes its first bite of the earth.

As the drill grinds its way into the earth, the friction of that grinding movement makes the bit very hot, and the *cuttings*, the bits of chewed-up earth from the well, can clutter up the hole, so a liquid called *drilling mud* or *drilling fluid* is used to cool and lubricate the bit and wash out the cuttings from the bottom of the hole. It isn't really mud, of course. It's a special kind of fluid made of water, oil, some clay, and chemicals.

A close-up of the rotary table on a production platform. When the table rotates, turning the kelly, the drill string is driven into the earth.

The mud is weighted in a special way, too, in order to add pressure to the drilling process. This mud is mixed up at the well site and stored there in the mud pit. During the drilling procedure, the mud is pumped from the pit through a hose called the *kelly hose* and through the swivel. The mud then travels through the kelly and drill string down into the well. Special jets in the drill bit force the mud out at the bottom of the well, where it collects all the cuttings and rock chips. The mud, with its cargo of cuttings, flows back up to the top of the well, through the space between the drill string itself and the *well casing*. Well casings are steel pipes that are lowered into the well as it is drilled in order to prevent its walls from collapsing. Once the mud reaches the top of the well again, the cuttings are removed and studied. By studying the cuttings, the geologist can tell what kinds of rock the drill is passing through. Meanwhile, the mud is cleaned of any debris from the well and used again.

After a few hundred yards of drilling, the bits wear out and must be replaced, so the driller makes a trip. The roughnecks use the lifting mechanism of the derrick, called the *draw works*, to hoist the drill string out of the well. Remember, by this time it could weigh 500,000 pounds! The drill string is broken into thribbles and stored in the derrick while the bit is changed. What comes up must go back down, so there are powerful brakes inside the draw works that can slow or stop the hoisting or lowering of the drill string. This may sound like a simple procedure, but remember that on a deep well, it can take fifteen hours to complete one trip.

As the bit grinds its way closer to the "pay sands," the rock layers that are likely to contain oil, the danger of a *blowout* increases. Oil and gas lying in their traps below the earth can be under incredible pressure. If the pressure inside the trap

Blowouts were common in the early days of the oil industry, and they could be very dangerous. This is a typical gusher at the famous Spindletop Oil Fields in Texas.

is too much when the bit finally pierces it, the oil or gas will run wild and blow up out of the hole, shooting pieces of drill string and drill bit into the air like bullets. A blowout (in the old days they were called gushers) can kill the workers and destroy the rig, cause a tremendous fire, or, at the very least, pollute the environment, so the driller carefully monitors the weight and pressure of the mud that is being pumped into the well. If the weight of the mud isn't heavy enough and the driller thinks that the oil in the well is tending to push toward the surface, he says the well is "kicking," and he stops the drilling process until the weight of the mud is adjusted.

The first blowout in the United States occurred at 10:30 A.M. on January 10, 1901 in Beaumont, Texas, in the Spindletop Oil Field. The well was 1,020 feet deep at the time. A fierce rumbling in the earth let the drillers know that something different was happening here. Suddenly a geyser of mud spouted up from the well, spewing drill string and pieces of the derrick high into the air. The mud was quickly followed by another rush of fluid—this time, oil. The Spindletop gusher continued to spew for nine days, creating an 800,000-barrel lake of crude oil around the well. In old movies, you often see people standing under a gusher of oil dancing around happily, as if it were a wonderful thing. After they ran for their lives, the Spindletop drillers were probably pretty happy, too. That well ushered in the petroleum age in America, and vast fortunes were made in the Spindletop Oil Field. Today, however, people are wiser. No one dances at a blowout. Everyone dreads one and tries to stop it before it ever starts. Fortunately, there is modern equipment today that helps the drillers prevent blowouts from happening.

A device called a *blowout preventer* is installed on every oil well. It is located in the *cellar*, a pit that has been dug below the production platform. At a station on the production platform, the driller monitors his instruments, instruments that tell him or her what the conditions inside the well are like and if the mud is heavy enough—or, if the pressure is building too rapidly, too heavy. If the driller decides that the pressure in-

Although there are modern methods to prevent blowouts, they still occur. This one took place in 1956.

side the well is too much and a blowout is likely, he or she will pull the hydraulic valves that activate the blowout preventer, and the well will be sealed off until the conditions can be corrected. Unfortunately, even with safety precautions and the best drillers in the world, blowouts do occur. Should that happen, the safety of the crew is another part of the driller's responsibility. The driller must be prepared to administer first aid in case of an emergency.

As the drilling continues and the drill bit moves into the rock layers that contain the oil traps, the well is "logged in." A log has many definitions. It can mean a large piece of wood,

but it can also mean a record of progress. The captains of ships and airplanes keep logs of their voyages. Men and women in the oil business keep logs on their oil wells, too. These logs record the progress of the drill bit as it bites and grinds its way through the layers of the earth. In order to log in the well, the driller makes a trip and removes the drill string and drill bit. Delicate instruments called *logging instruments*

The Spindletop gusher marked the beginning of the petroleum age. Soon there were forests of wooden oil derricks crowding the plains around Beaumont, Texas.

Above: A blowout preventer is repaired for a drill ship.
Right: Work continues at night on a rig in Alaska.

are lowered into the well. The logging instruments give oil drillers a picture of what the inside of the well is like. For instance, they may want to know something about the kind of rock formations they are drilling into. They will want to know something about the *porosity* of the sandstone. In other words, how big are the holes in the rocks? Porosity affects how much oil could be trapped there. They also would like to know some-

thing about the *permeability* of the rock layers they are penetrating. To permeate means to pass through, and when oil people discuss permeability, they are talking about how easy it will be for the oil to flow from the rock formations where it is trapped into the bottom of the well. That kind of information lets them know if a blowout could occur when the oil trap is pierced. If it could, they will take action to prevent it.

The last step in drilling an oil well is cementing. The steel well casing is centered in the hole, and cement is pumped down through the middle of the casing, out through the bottom, and up between the sides of the casing and the hole. The cement seals off the well and strengthens it against a blowout. A logging instrument is lowered into the well after the cementing is complete. It will check the seal between the cement and the sides of the well, letting the driller know if the well is secure. If everything is all right, the cement at the bottom of the well is perforated, or pierced, and the oil is released from the trap where it has been held for the last 250,000 years or so.

When the well is ready to go into production, a set of valves and fittings called a *Christmas tree* is put into place. The first Christmas tree was designed by Al Hamill, a driller on the Spindletop gusher, to stop the gusher that was flooding the surrounding area with oil. (If you use your imagination, you can see that the collection of valves and pipes looks a bit like a skinny fir tree decorated with ornaments.) Today's Christmas trees are used to control how quickly the oil will be pumped to the surface, where it is stored before it begins the long journey to the refinery to be changed into the products that you and I use every day.

A Hole in the Bottom of the Sea 2

It is much easier, and far less expensive, to drill for oil on land; but that oil is getting scarce. More and more often, we look to the oil that is buried beneath the floor of the sea to supply our energy needs. Drilling for that oil, while similar in some ways to drilling on dry land, is a far more dangerous and difficult proposition.

When an oil company thinks that oil might be trapped beneath the waters of the sea, it must move its drilling operation onto barges or drilling ships or offshore drilling rigs. A drilling operation at sea requires tool pushers, drillers, derrickmen, and roustabouts, just as it does on land. But they must live on board the barges, ships, and offshore rigs for weeks at a time, far away from their homes and families. Also on board are other people, such as geologists and deep-sea divers, launderers and caterers.

Helicopters and supply boats go back and forth, transferring crews, support personnel, and supplies from land to sea. Some of the most important people are the caterers, who stay on board to cook the meals. Because there is little to do for entertainment when one is working offshore, the food is particularly important. Some of the best food in the world is served to the crews of offshore oil operations.

The tours at sea usually last twelve hours at a time, instead of the eight-hour work shifts on land. Some crews work seven days in a row and then come ashore to be with their families

The search for oil continues on offshore drilling ships.

for seven days. Other crews work fourteen days out, then have fourteen days at home. The time at sea can be boring when the people on the crew are not actually working. There is no local bowling alley or movie theater to visit. Television signals can't reach the offshore drilling operations, either. Nevertheless, during time off there are videotaped movies to watch, video and other kinds of games to play, books and magazines to read, and letters to write before the tired crew goes to bed.

The drilling procedure at sea is the same as it is on land. Using derricks and drill strings, drill bits, drilling mud, and cement, the drilling crew sinks its hole deep into the ocean's floor. The average depth of a well on land is 5,000 feet. At sea, the bit may have to drop through hundreds of fathoms of water before it even begins to chew into the ocean's floor. All over the waters of the world, from the decks of barges, drilling ships, jack-up rigs, and giant semisubmersible platforms, people are searching for the oil and gas that we all need. Let's take a look at each kind of offshore operation.

When an oil company decides to drill a well where the water is relatively shallow, such as a bay or river, it might choose to drill from the deck of a barge. A barge has a shallow hull, and it doesn't need as much water to float as, say, a drilling ship. Barges do not have engines, so tugboats pull them into place, and they are anchored there for the length of time it takes to drill the well. All the equipment for drilling the well, along with the housing for the crew, is mounted on the deck of the barge. The crew lives and works on this anchored island until the well is completed. Then the tugboat returns, the anchors are lifted, and the barge is taken to its next location. The barge will always drill in shallow water, because it does not have the stability to remain steady in the rough waves and currents of deeper water. If an oil company wants to drill in

A drilling ship moves toward the oil fields beneath the sea. The target on the ship's stern marks the helicopter platform, which is standard equipment on the ships.

the deep water out at sea, it may decide to use a drilling ship.

A drilling ship looks a bit like an ocean-going freighter, except, of course, for the fact that a derrick is sticking up right in the middle. A helicopter platform is standard equipment on the stern, or rear, of the ship. The derrick is mounted over the *moonpool*, which is the name of the hole that has been cut through the ship's hull so the drill can plunge into the water. Unlike a barge, a drilling ship is self-propelled, so it requires two kinds of crews—one to operate the ship and one to drill the oil well. It is more expensive to drill an oil well from the deck of a drilling ship than it is to drill one from a barge, but a drilling ship has the advantage of being able to work in much deeper water than a barge is able to do.

Not all offshore drilling is done from the deck of a barge or drilling ship. If you were to stand at night on the beach at Galveston, an island just off the coast of Texas, and look out to sea, you would see dozens and dozens of brightly lit structures dotting the horizon. Some of these are *jack-up rigs*, and they are well suited for working in shallow coastal waters. These rigs, which look a bit like a giant table with many legs, are built on the land, then towed to the well site with their legs in the raised, or "up," position. Once the rig gets to its destination, the legs are lowered to the ocean's floor, and the rig stays put. The table's "top," the actual drilling platform, sits high and dry above the surface of the sea, providing a man-made island for the men who drill for oil and their machines.

So far we have discussed drilling operations in water that, for the most part, is relatively calm. However, there are rich deposits of oil under the waters of the North Sea and under the Gulf of Alaska in the frigid Arctic. The weather conditions in both those places can be terrible. At times the winds howl and blow at speeds of over one hundred miles an hour. In places, the water is 4,500 feet deep. Waves eighty to one hundred feet high crash and pound. In the Gulf of Alaska, the temperature can drop to $-30°F$. Very specialized drilling rigs are used in these kinds of conditions. They are called *semisubmersible rigs*, and they are custom-built for the waters they will be drilling in. A semi (short for semisubmersible) has its drilling platform mounted on top of great, fat legs, or columns. Like the barges, drilling ships, and jack-up rigs, the semis have heliports and living quarters as standard equipment. Some of these rigs can house a crew of over one hundred. The rigs are constructed on land, then moved—either under their own power or by tug—to the well site. Once a rig is there, the great legs are flooded with water, allowing them to sink to the ocean's floor. However, a semi's legs do not have to rest on the bottom of the sea; the rig can be submerged to any depth the oil company wants, then held in place by anchors and cables. Workers called *ballast men* are responsible for making certain that the correct amount of water is in each leg so that the rig

Above: A semisubmersible rig is a home at sea for many workers.
Left: Three tugboats tow a jackup rig to its drilling site.

Tugs tow this 348,000-ton production platform on a 250-mile voyage to the North Sea. The legs are in the up position until the platform reaches the drilling site.

stays level. A semisubmersible rig is not a permanent installation. The amount of water, or ballast, in the columns can be changed, so the columns will lift from the ocean's floor, allowing the rig to be floated to a new location as the need arises.

Once oil has been discovered offshore and the oil company decides that there is enough good-quality oil in the field to be worth producing, another kind of offshore equipment—the floating production platform—is needed. These giant platforms are some of the largest moving structures that man has ever built. The Ninian Central is an enormous production platform that sits in the Ninian Oil Field in the North Sea. Several hundred feet taller than the Washington Monument and weighing over 600,000 *tons*, it provides a home away from home for over two hundred oil-field workers. Production platforms like this cost millions of dollars to construct, so oil companies drill several oil wells from one platform. The Ninian Central is one of many giant production platforms that dot the North Sea and other bodies of water. It is owned by Chevron Oil Company, and twenty-seven producing oil wells have been drilled from its surface alone.

The oil companies take every precaution to ensure the safety of the people who work aboard their rigs and production platforms. Medical supplies are always on hand, and the rigs and platforms are inspected regularly to make sure there are plenty of fire extinguishers and pieces of lifesaving equipment on board. However, accidents can, and do, happen. The decks are slippery, and sometimes when the waves crash and the winds howl, workers can be washed overboard. On March 27, 1980, a storm blew up in the North Sea. The winds quickly reached 65 miles per hour, and waves rose 30 feet in the air. The fierce storm hit unexpectedly, and its force snapped one of the legs of the Alexander L. Keilland, a floating production platform. The crippled platform tilted over at a 45 degree angle. Twenty minutes later, it capsized. There were 212 workers aboard that platform when the storm hit, and 123 of them were washed overboard and died. The tremendously high seas made it impossible to launch all seven of the 50-man lifeboats that the platform carried. However, some lifeboats were launched and 89 people were eventually rescued by search teams from five different nations. Besides natural

disasters such as storms, there is always the danger of a blowout, even though blowout preventers are installed on the wells at sea, just as they are on the wells on land. Should a blowout at sea happen, fire could destroy the rig even before it sank. Blowouts are rare, but nevertheless, watertight escape capsules for the crew are often part of a rig's standard equipment.

On some rigs, deep-sea divers are part of the support personnel. It is their job to dive into the waters to inspect and, if necessary, repair the underwater parts of the well. Some rigs have decompression chambers on board to prevent divers from getting the bends, a popular term for a disease that causes painful, paralyzing, sometimes fatal, contractions of the body as it tries to adjust to the difference in pressure between under the sea and at sea level. It can take a long time for the body to decompress, and on some offshore rigs, the divers spend all of their time in the decompression chamber when they are not actually diving.

Life on an offshore drilling rig, or a production platform, is always noisy, dangerous, and difficult. When the crews aren't working, it can also be boring, but the work is important, and the pay is very good. Counting overtime and extra pay for holidays, some offshore oil workers can make as much as $6,000 a week.

So far we have talked about the dangers to the people who search for oil in the waters of the world, but there is danger to the environment, too. A blowout on an offshore well can cause an oil spill that damages beaches and kills fish and birds for hundreds and hundreds of miles. In 1979, there was a blow-

Seas can get very rough around the North Sea platforms.

out on a well off the coast of Campeche, Mexico. The fire from that blowout burned for months. Companies who specialize in fighting oil well fires were called in to attempt to put out the blaze—a job that can cost millions of dollars and take weeks, sometimes even months, to complete. The oil companies that

Above: Work on the oil well can be difficult and dangerous.
Left: Deep-sea divers keep the well in good condition.

Seals take a sunbath on a production platform mooring buoy.

own the well must also pay for the clean-up operations when a blowout occurs. It is hard to mop up oil that has been spilled in the sea. One way to try is by stringing long nets, with floats on the top and weights on the bottom, in the water, similar to seining for fish. However, in this case, instead of catching fish, the nets catch the gummy globs of oil before they reach the beaches. Nevertheless, it is often impossible to catch all of the oil, and in the 1979 blowout in Mexico, beaches as far north as Padre Island, Texas, were ruined for quite a while because they were covered with tarlike oil deposits that escaped the nets and floated to shore, killing fish and water birds along the way.

On a happier note, there is some evidence that the offshore oil rigs actually increase the numbers of fish in the sea. The drilling mud that is pumped into the sea forms an artificial reef on the bottom. The reef attracts fish and other marine life, so many deep-sea fishermen in the Gulf of Mexico prefer to sink their hooks close to an offshore oil rig. Why? The fishing's better.

Often the oil needs to be pumped from the well. Most pumps are black, though these have some patriotic decorations.

Bringing It In 3

Whether the crew is drilling on land or far out to sea, the moment comes when its job is over. The well has been logged in, and the Christmas tree is in place. By this time, the derrick, drill strings, and drill bits have disappeared, along with the tool pushers and drillers, roustabouts and derrickmen. Their job is finished, and they move on to drill another well in another field. The oil now becomes the responsibility of a person called the *pumper*.

As you have seen, the Christmas tree regulates the flow of oil coming from the well. Some oil—the kind that produces a gusher—is under tremendous pressure and will spurt to the top of the well if it is not controlled. Other deposits of oil will flow to the top of the well easily and naturally. But most deposits will need a boost from a pump in order to get to the top. If you drive by an oil field, you will see pumps nodding their heads up and down like lazy horses grazing in a meadow. As they pump the oil from the ground, the pumper is there to see to it that the oil is stored properly in the tanks at the well site.

Many times an oil well contains deposits of natural gas as well. Gas is lighter than oil, so in the trap it floats on top of the oil. If salt water has been trapped too, it will float on top of the gas, because it is lighter than both gas and oil. These three elements—salt water, gas, and oil—must be separated from one another at the well site. The fluid is pumped from the well directly to a tank called a *gun barrel tank,* and indeed, because it is thinner and taller than most of the other, rather squat, storage tanks at the well site, it *does* look a bit like a gun barrel. The gun barrel tank is a settling tank that allows the oil to settle on the bottom while the gas and salt water float on top. The pumper then separates the gas from the oil. The gas goes into one tank, and the oil and salt water go into another. The salt water and other debris—bits of sand and mud—are removed from the oil at that point. It is the pumper's job to see that the amounts of gas and oil that are pumped from the well are measured and that those measurements are recorded. Individuals, companies, or governments own the land that the oil well is on, and they must be paid for the oil and gas that come from the well. The payment is called a *royalty.* A royalty is a fee for each barrel of oil or cubit foot of gas that is pumped from the well. The pumper keeps track of those barrels and sees to it that the owner receives the proper credit.

Oil straight from the well is called *crude oil,* sometimes just crude for short. While the crude waits in its storage tank for the trip to the refinery, a person called a *gager* uses an instrument called a *tank thief* to sample it and decide whether it is heavy or light crude. Light crude is, as its name implies, a lighter kind of oil. It is thinner and flows more freely than heavy crude. Light crude produces the most gasoline, and so is considered the most desirable crude and usually brings the highest price per barrel. At the well, crude oil is priced by the barrel. (A barrel of oil contains forty-two gallons.) By the time you step into a store to buy a small can of lubricating oil, it has been priced by the ounce! Heavy crude is thicker and—because of the sulfur in it—smellier than light crude, but plastic, paint, asphalt, vinyl, and hundreds of other products, in addition to gasoline, can be refined from heavy crude oil, so a barrel of heavy crude is very valuable, too.

Before going to the refinery, the crude oil waits in storage tanks at the drill site, whether that site is on land or out at sea. The tanks that gather the offshore oil are built on the land, then taken to the well site and flooded with water. The water-filled tanks sink to the bottom of the sea, waiting for the oil from the well to flow into them. As they fill, the oil pushes the water inside them out, so they are always full of *something* and there is no danger that they will bob up from the ocean's floor. On land or sea, when the tanks are full, the oil is ready to begin its trip to the refinery, where it will be turned into products that you and I use every day.

Since the industry began, pipelines have been used to carry oil. The pipelines that serviced these wells in 1870 were wooden boxes.

Pigs and Pipelines, Trucks and Tankers

The first people to move oil or gas from the well to another location were the Chinese, as we mentioned in the first chapter of this book. They used pieces of bamboo to pipe natural gas from the well to the place they were going to use it. Today, pipelines carry crude oil from the storage tanks at the well either directly to the refineries or to the barges, tankers, and trucks that will deliver the oil to the refineries. Some of these pipelines are buried in the earth, about one yard deep in most cases; others rest on river and ocean bottoms; still others lie on top of the earth.

The longest, most famous, and most expensive pipeline in the world is the Alaskan Pipeline. It begins at Prudhoe Bay, Alaska, and snakes across three mountain ranges and through more than two hundred rivers before it ends, eight hundred miles later, at the port of Valdez in southern Alaska, where it is loaded into tankers to be shipped to refineries along the Pacific coast. It was finally finished in 1977, and it cost more than 7 *billion* dollars to build. Some 1.5 million barrels of crude oil travel through it each day.

When plans to build the Alaskan Pipeline were made, there were those who thought it should not be built. They were worried about damage to the environment and to the ecology in that part of the country. (Ecology refers to the relationship between living things and their environment.) Much of the land in Alaska is composed of *permafrost* and *tundra*. Permafrost is exactly what its name implies—land that is permanently frozen year round. However, a temperature change of one degree could cause the permafrost to begin to melt, and then, when the ground refreezes, it buckles upward. The oil must be kept warm so it will flow through the pipeline, but it can't be too warm or it will melt the permafrost, causing a buckling problem.

Tundra is a sort of frozen swamp, and Arctic wildlife feeds on its sparse grasses. The soil in the tundra is very delicate and can be easily damaged. Once damaged, it is impossible to repair. Damage to the tundra can also upset the ecology of the area, causing the unnecessary death of plants and animals. In order to protect both the permafrost and the tundra, the pipeline is built on supports above the ground as it runs across these areas.

Above: Drilling rig in the frozen wastes of Alaska.
Left: The Alaskan Pipeline, the world's largest.

The Alaskan Pipeline, and others too, run over many different kinds of terrain. Because pipelines can climb mountains and descend into valleys, you can see that the oil in them does not necessarily run downhill all the time. Every once in a while along the way, there is a pumping station attached to the pipeline. The pumping station pushes the oil inside the pipeline uphill and down, toward its destination—a port, a refinery, or a manufacturing company.

A complete pipeline system has three different kinds of pipelines in it—gathering lines, trunk lines, and distribution lines. This system works a bit like the circulation system in your body. Your blood is sent to and from your heart and lungs (which we can compare here to the refinery) by a network of veins and arteries. The veins that gather the blood and take it to the heart start out small and gradually get larger as they accumulate more and more blood from different parts of your body. The arteries send out the blood that has been pumped through your heart and lungs. The arteries reverse the process, starting out large as they send all your oxygenated blood on its way and gradually becoming smaller and smaller as the blood finally arrives at your toes and fingertips.

Oil gets to the refinery in much the same way. The *gathering lines*, which pick it up from the faraway oil wells in the field, are small. Eventually, this network of small pipelines feeds into a larger pipeline, the *trunk line*, which carries the crude to the refinery, or perhaps to a tanker that will take it to the refinery. When the refining process is completed, the situation is reversed, just as it is in your body. The refined products are pumped into *distribution lines*, which start out large and gradually become smaller as the various products are delivered to their destinations. Depending on how many wells or refineries they are servicing, gathering and distributing pipelines can be anywhere from 4 to 12 inches in diameter.

Special barges lay pipelines under the sea.

Oil companies share the use of pipelines, so oil is run through them in batches, whether it is coming in from the field in the gathering part of the system or leaving the refinery by the distributing part of the system. The oil batches are separated by sending water or kerosene through the pipeline between batches. Sometimes dye is run between batches so that each company's oil can be identified.

A pipeline system is made of lengths of steel pipe that are welded together, then covered with a special protective coating, usually plastic if the pipeline runs under or above the ground, concrete if it's sunk beneath the sea. It is very important that the pipeline not leak, so it is inspected regularly.

Inspectors fly over the pipeline in helicopters if it is running above the ground in remote locations. In easy-to-reach locations, they simply walk along beside it, looking for possible leaks or signs of wear. Pipelines beneath the sea must be inspected, too, and deep-sea divers are called in for this task.

In addition to being regularly inspected for leaks, the pipeline must be kept clean. Anything that has oil running through it constantly is bound to get dirty! Appropriately enough, a device called a *pig* is used for this kind of work. The pig is pushed through the pipeline ahead of a new shipment of oil. As the pig travels through, it scrapes the sides of the pipeline, removing build-ups of dirt and wax that have accumulated there. The pig can be removed at pumping stations so that it, too, can be cleaned and returned to its job.

If an oil company is small or if an oil field is located in a very remote area, the wells may not be connected to a pipeline system. In that case, the oil company may choose to move its product to the refinery by truck. You have probably seen these tanker trucks whizzing along the highway. Trucks also transport finished products, such as gasoline and jet fuel, and deliver them to the waiting service stations and airports. Sometimes the tanker trucks pick up the products at the refinery itself, and sometimes they pick up the product at the end of the distribution pipeline.

Railroad transportation, the first method of moving crude oil, was most popular until the time of World War II. Then, the large amounts of petroleum needed to support the war effort were shipped on oil tankers that sailed along America's coasts, taking oil, for example, from Texas oil fields to refineries along the east coast. However, many people felt that the coastal tankers were too vulnerable and could be easily sunk by an enemy submarine (in fact, off the coast of New

A delivery salesman loads his tank truck at a pipeline terminal in Dallas, Texas. The gasoline that has been made from crude oil is on its way to the customer.

Jersey, several were!), and they began to look for other, safer methods of transporting the oil. One solution was the construction of two new pipelines—Big Inch, which was 22 inches in diameter, and its 20-inch counterpart, Little Inch. The Big Inch and Little Inch pipelines run from Texas to the east coast, delivering oil efficiently and safely today, just as they did forty years ago.

Our nation's rivers and the Great Lakes have been used to transport oil and other products for years, but rivers flow from the north to the south, and the oil needs to be sent east and west, too. Pipelines and trucks and railroad tank cars handled some of this job, but they were not enough to move the vast amounts of oil needed during World War II. So, at that time, in addition to the pipelines, the Army Corps of Engineers built miles and miles of inland waterways and canals that crossed

A special scrubbing machine is used to clean the tanker's hull.

Oil tankers provide the cheapest and most efficient way to transport oil. This tanker is docked in Linden, New Jersey. It can load or unload oil from dockside.

the nation running east and west. Barges pushed by towboats began slipping along these inland waterways, taking oil to its destination safely and inexpensively. Although oil is still transported to the refineries by rail and by tanker truck, getting America's crude oil to America's refineries is handled more and more by barges towed along our inland waterways.

However, the cheapest and most efficient way to get oil to its destination around the world is not by barge or by rail or by tanker truck. Today, the cheapest and most efficient method of transporting oil is by sea in the cargo hold of an oil tanker.

Oil tankers come in many sizes. The smaller tankers work in coastal waters and do not cross great oceans. They are used to deliver oil from Houston, Texas, to Bayonne, New Jersey, for example. The tanker is docked at port, and the oil is loaded

This 500,000-ton ULCC dwarfs the VLCC that floats by its side. It can carry almost twice as much crude oil.

directly into the ship by means of a pipeline. But if the oil is going from Houston to a location in Asia or Africa or Europe, it is loaded into the holds of huge oil tankers called *VLCC*'s, for very large crude carriers, and *ULCC*'s, for ultra large crude carriers, while they are floating at anchor out at sea.

A VLCC can carry 276,000 *tons* of crude oil, and her big sister, the ULCC, can carry over 500,000 tons. The more crude a ship can carry, the less expensive it is to move it. These ships, which are little more than floating oil tanks, are the giants of the sea, dwarfing every other kind of ocean-going vessel. They can be 1,300 feet long and 200 feet wide, so big that the crew uses bicycles to get from the bow, or front, of the ship to the stern.

Surprisingly enough, it takes a crew of only about forty men to sail these mammoth ships, because most of the sailing is done automatically. Sophisticated instruments and computers keep the boat on course when it is crossing the world's waters. On board, the ship's radar equipment warns of other ships in the area, and sonar equipment keeps tabs on the depth of the water at all times—a ULCC needs at least 95 feet of water just to float! The crew must, of course, monitor the instruments and know how to handle the giant ship in case of emergency.

The most dangerous times on board a large tanker come when it is loading or unloading its cargo of oil. Because they are so big, these dinosaurs of the sea are extremely difficult to maneuver in a small space. They travel at speeds of about twenty knots an hour, and even going that slowly, it may take the captain at least twenty minutes to stop—he may still travel a few miles after he has put on the brakes! Actually, there are no brakes on oil tankers, just as there are no brakes on airplanes. In order to stop, the captain must reverse the engines.

To do this, he turns the propeller in the opposite direction, literally backing the ship up until its forward momentum stops.

A VLCC or ULCC could never dock in a regular port to load or unload its cargo. The water is simply not deep enough. And because they are so difficult to handle, they cannot be pulled up alongside an offshore drilling platform, either. If they were, they might swing into the platform and send it tumbling into the sea like a house made of matchsticks.

While picking up or unloading oil at sea, the huge tankers

A tanker is taking on a cargo of crude oil from a floating storage buoy. A single line tethers it to the buoy.

must dock at a device called an *SBM*. These letters stand for single buoy mooring. The SBM is anchored at sea in deep water, and the tanker is tethered (sea talk for "tied") to the buoy by a single line, allowing it to float around the buoy freely while different compartments in its insides are being filled with or emptied of crude oil. Oil from storage tanks on land or under the sea is pumped to the SBM by an undersea pipeline. The oil is then loaded into the many cargo holds of the tanker through a flexible pipe that floats on the water. A tanker unloads its oil by the same process, sending it to the buoy via the flexible pipe, then from the buoy to the storage tanks by undersea pipeline.

There are good reasons for loading the oil into separate compartments inside the tanker. For one thing, a large tanker may deliver oil for several different companies, and naturally, the oil has to be separated, just as it is in the pipeline. The different compartments in the ship do this job nicely. Then too, even if all the oil belonged to one company, if it were allowed to slosh around in one big batch inside the hull, it would throw the ship off balance when it was out at sea. The compartments keep the oil evenly distributed. Also, if an accident such as a collision or running aground should occur and one of the cargo holds broke, at least only the oil in that hold, not the entire cargo, would be spilled into the sea. The giant tankers unload their cargo at deep-sea terminals, and from there it is transported directly to the refineries, either by pipeline or by rail or truck tank cars.

There are people who feel that the VLCC's and ULCC's, because they are so big, are dangerous. Because they take so long to stop and turn, they do have more collisions than other kinds of ocean-going ships. And because their draught (the part of the boat that sinks below the water) is so deep, it is easier for them to hit a shoal, ripping a hole in the hull. People remember incidents like the time in 1976 when a Liberian tanker named the *Argo Merchant* went aground on the shoals off of Nantucket, Massachusetts. Five million tons of oil spilled into the Atlantic Ocean in that one accident. The beaches all along the New England coast were in danger, and so were the fish and water birds. In 1979, there was an explosion and fire aboard the French oil tanker, *Betelgeuse*. She was unloading her cargo in Ireland when double explosions blew the tanker in two pieces. Fifty people were killed in that tragedy.

On the other hand, the people who support the big tankers argue that they keep the cost of oil down, which is true. VLCC's and ULCC's are the cheapest way of transporting oil around the world, and using them means lower oil and gasoline prices for everyone. Their supporters also argue that because the tankers hold so much oil, it takes fewer of them to get the cargo to its destination. They say that with fewer ships with this kind of explosive cargo at sea, there is really less chance of a collision than if more, small tankers were being used.

Life on board a large tanker is similar to life on an offshore oil rig, except the crew of a tanker is often away from home for months, instead of weeks, at a time. The control room and the crew's quarters are usually at the stern of the ship, as far away from the fumes and the danger of fire as possible. On very large tankers, members of the crew have their own cabins. As on an oil rig, the food is very good, and there are books and movies and video games for entertainment. Some tankers even have a swimming pool! Occasionally, the captain and other officers are permitted to take their families out to sea with them, but most of the time the voyage is a lonely one.

During the Christmas season, many people in Houston,

The deck of this large tanker is as big as many city parks. Crew members use bicycles to travel from the bow to the stern.

Texas, fix gift boxes for the crews of the tankers and other ships that may be in port over the holiday season. They fill shoe boxes with bars of soap and shaving lotion, shoelaces, notepaper, stamps, candy, and anything else they can think of. Often they will enclose a picture of their family with a letter that tells something of what it is like to live in Houston. Some of these families have in return received letters and pictures from the crew on the tankers, telling what life at sea and in their home country is like.

The refinery is the last stop for a batch of crude oil.

Boiling the Oil

At night, lit by thousands of lights, flames burning at the top of its many towers, a refinery looks a bit like a space station that somehow has been grounded on earth. This maze of towers and tanks, catwalks and control rooms, is the last stop for a batch of crude oil that could have begun its journey here from half a world away—or from an oil storage tank only miles down the road. Crude oil comes to the refinery to be purified and changed into other products, such as gasoline for lawn mowers and cars, or paraffin to seal homemade jelly and jam.

Crude oil wasn't always refined. Often it was sold just as it came from the well, for use as a medicine or a grease. In the 1840s, a Scotsman named James Young began experimenting to see if any other products result when coal is heated. He found that he could get something called *coal oil* from this process. Coal oil is also called *kerosene*, and Young, who was quite an inventor, built a lamp that would burn the coal oil. Soon kerosene lamps were taking the place of candles for people around the world. About ten years later, in the 1850s, Samuel Kier was wondering what to do with all of the oil that was puddling up on his Pennsylvania farm. He wondered if he could get it to produce coal oil, or kerosene. So he heated his crude oil just as Young had heated his coal, and sure enough, kerosene resulted. Kier bottled his kerosene and sold it as a medicine called Kier's Rock Oil. Kier's Rock Oil was sold as a panacea, a cure-all medicine that was, in the words of the old-time traveling medicine shows, "good for what ails you."

Another man in Pennsylvania, George Bissell, was interested in producing kerosene from his crude oil, too. He sent a sample of the crude from his farm to Professor Benjamin Silliman, a brilliant chemist at Yale University. In his laboratory Professor Silliman heated Bissell's crude oil to temperatures as high as 750°F. At this temperature, he noticed that kerosene wasn't the only product that could be had from crude oil. It also produced a waxy substance called *paraffin* and heavy lubricating grease. Professor Silliman sent his reports to Bissell, and Bissell drilled the first commercial oil well in the United States.

The first refining process used heat. The crude oil was *distilled*—that is, heated until it turns into a vapor, just as water, when it is boiled, turns into steam. Water boils and turns into a vapor at a temperature of 212°F., but other liquids boil and vaporize at higher or lower temperatures. If you watch a pot of water come to what cooks call a rolling boil, you will see that bubbles constantly form on the surface and burst, turning into steam, which, when it hits the cooler air of the room, condenses back into a liquid and forms the water drops that sometimes settle on the cabinets and counter tops near the stove. The distilled water in the drops is different—changed—from the water that you put in the pan to boil. It is now distilled water. Because it has been turned to steam, then condensed back into liquid, the water is purified. Oil is refined, or changed, in the same manner.

The refining process begins by pumping the crude oil into a furnace near the base of a tall tower, called a *fractionating tower*, where crude is separated into various products. In the furnace, the oil is heated to around 800°F. and is then pumped to the bottom of the tower. The temperature in the tower is

Crude oil is separated in the fractionating towers.

hottest at the bottom (around 700°F.) and coolest at the top (a chilly 104°F.!). However, even with that wide difference in temperature, the tower is filled from the top to the bottom with boiling *fractions* of crude oil. A fraction is a part of something else, so fraction is a good name for the gases, gasolines, kerosenes, heavy oils, and asphalts that are part of, and refined out of, crude oil.

We have seen how distilled water is separated from regular tap water by boiling it. The fractions of crude oil can be separated from one another by boiling, too, because each fraction boils and vaporizes at a different temperature. For example, crude oil *begins* to vaporize at 104°F. The vapors that form at around this temperature are of the lightest fractions of crude oil, and they condense at the top of the fractionating tower, where the temperature is coolest. As these light vapors condense, they turn into liquid gas, formally called *liquified petroleum gas*, or LPG for short. This kind of gas is used for some cooking and camping equipment, for example, some camping stoves.

A little farther down in the fractionating tower, where the temperature is running about 356°F., the part of crude that will turn into gasoline is boiling, vaporizing, and condensing. This is the fraction everyone is so anxious to get, because gasoline powers our cars and lawn mowers, go-carts, and mopeds—anything that runs with a gasoline-powered motor. Without gasoline, our society would literally come to a stop!

Lower down in the tower, it is up to 446°F. The product, or fraction, that is produced here is kerosene—the very product that started the demand for oil in the first place. Today, there's not much demand for kerosene lamps, but kerosene is the major ingredient in jet-airplane fuels, so we could not do without it.

Still lower in the tower, the heavy oils are refined out of the crude oil. It takes a higher temperature to make them boil. Diesel fuel, the kind that runs trains, eighteen-wheel trucks, and some cars is distilled and collected here at a temperature of 581°F.

Closer to the bottom of the tower, the temperature is even hotter. Heavy oils are finally boiling and vaporizing here at temperatures of 761°F. These oils are called *lubricating stock*, which provides the kind of oil you use to grease a wheel or loosen rusty nuts and bolts. You may have a jar of petroleum jelly in your medicine cabinet at home. It is made from the lubricating stock fraction, and it is good for soothing blisters and diaper rash, cuts and burns.

Residuum collects at the bottom of the fractionating tower. The term comes from the word residue, which means something that remains after other things have been taken away. The residuum remains in the bottom of the fractionating tower after all the other products have condensed from the crude oil. It is a sticky black mass that is often turned into asphalt to pave roads and driveways and schoolyards.

The trays in a fractionating tower that collect the fractions are not solid. They have holes in them so that the vapors can pass through the lower trays and on up to the top of the tower. The holes in the tray have tops called *bubble caps* suspended over them. Bubble caps allow the crude oil to pass through the trays as a vapor; then, when it condenses again to a liquid, the bubble caps force the liquid fraction to run off to the sides of the trays, where it is collected for further refining. Inside the fractionating tower, all of this activity is taking place at once. The various fractions of crude oil are bubbling, vaporizing, and condensing at their different temperatures—and heights—up and down the tower.

Finally, in each barrel of crude oil that is refined, there is some fraction that is so light it simply won't condense into a liquid at all. It remains as a gas and is burned off at the top of the fractionating tower. At night you can see these torchlike flames burning on top of the refinery's tower.

The various fractions we have discussed are drawn from their trays in the fractionating towers and go through many other processes before they leave the refinery. You should know about two of these processes—one is called *thermal cracking* and the other is called *catalytic cracking*.

As we have said, for years gasoline has been the fraction of crude oil that everyone wants. The problem is, under the dis-

tillation method of refining that we have just discussed, too many of the heavier petroleum products and not enough gasoline were produced. So chemists and engineers began to try to find a way to get a barrel of crude oil to yield more gasoline. One of the people who worked on this problem was Dr. William Burton, a chemist with Standard Oil of Indiana. Dr. Burton believed he could break up, or crack, the heavy residuum in the bottom of the fractionating towers and get it to yield more gasoline. He put the residuum in a heavy vessel

Natural gas is processed on the King Ranch in Texas.

called a *still* and applied heat and pressure, and sure enough, he managed to produce gasoline from the fraction that everyone had thought could be used only for asphalt or wax. This process of applying heat and pressure to the residuum is called thermal cracking, and it is still used, with variations and improvements, today.

One of the men who improved on Burton's thermal cracking process was a fellow whose parents must have wanted him to go into the oil business, because they named him Carbon Petroleum Dubbs. Dubbs knew that one of the problems with Burton's method of thermal cracking was the dirt and residue that remained in the still after the residuum was cracked and the gasoline removed. The still could be run for only a few hours before it had to be shut down for cleaning. As in all businesses, time is valuable, and the Burton process wasted time. Dubbs's method, which he called the Dubbs Process, produced much less residue, so the cracking process could go on for days at a time before cleaning was necessary. It was a major improvement in thermal cracking, and the Dubbs Process is still used by refineries today.

Another method of breaking down the heavy-oil fractions to get them to produce lighter oils is catalytic cracking. In catalytic cracking, the hot vapors of the heavy oil are pumped into a heavy, pressurized tank called a *reactor*. Then a substance called a *catalyst* is added. When a catalyst is added to something, it causes a chemical change to happen. During catalytic cracking, the catalyst acts on the atoms and molecules in the oil, changing their composition and turning the heavy oil into the desired lighter fractions. During this process the catalyst itself never changes. Different catalysts can be used to start this process, and they are cleaned and used over and over again.

In 1872 a brilliant Russian chemist named Dmitri Mendeleev visited Pennsylvania. He was interested in the crude oil that was produced there. Mendeleev knew that crude oil is composed of elements of hydrogen and carbon—hydrocarbons, for short. He also knew it contains sulfur and nitrogen. After his time in Pennsylvania studying the oil operations there, he reported back to his government. He said that, in his opinion, oil is too valuable to burn as a fuel at all and should be used only as a chemical base material.

As the refining process grew more and more sophisticated, scientists like Mendeleev agreed that the four elements—hydrogen, carbon, sulfur, and nitrogen—that appear in crude oil could be rearranged chemically and turned into substances called *petrochemicals*, which is simply another way of saying chemicals made from petroleum, the base material Mendeleev was talking about. If Mendeleev were alive today, he would be pleased to see the growth of the petrochemical industry, because that industry turns oil into the base material that provides thousands of products that each of us uses every day. You hear their tongue-twisting names—methane, propylene, ethylene, butylene, naphthene—all the time in television commercials. They are used to make products as varied as deodorant and dishes, mouthwash and aspirin, football helmets and food coloring, the ink in this book, the Orlon in your sweater, the plastic in your comb, the tires on your bike, explosives, and anesthetics and drugs. All of these and more are made from a fraction of crude oil. Without petrochemicals, our lives would change drastically.

Although an oil refinery looks huge and its oil storage tanks, fractionating towers, pipelines, and catalytic cracking units cover hundreds of acres of land, it takes relatively few people to operate one. Like the tankers, oil refineries are run largely by automation. About forty people work in a control room supervising the instruments that monitor the pressure of the crude oil as it passes through the fractionating towers, pipelines, and catalytic cracking units. The possibility of explosion and fire is always present and always a danger, so most refineries have their own fire departments and first-aid stations right on the premises.

The waste products from refining oil must be cleaned before they are released into the air. Sulfur is a valuable element in crude oil used in the manufacture of petrochemicals. However, removing sulfur in the refining process produces a gas called sulfur dioxide. Sulfur dioxide has an extremely unpleasant odor. In fact, it is often compared to the smell of rotten eggs. Besides making it unpleasant to breathe, too much sulfur dioxide will actually pollute the air around a refinery and cause a health problem for the people who live and work in the area.

Oil Gluts and Shortages and the Energy Crisis 6

Energy crisis—the term is familiar to all of us. But exactly what *is* an energy crisis, and do we really have one? The answer to the second question is both yes and no. Yes, we do have a crisis, because today, 78 percent of the energy needs of the United States are met with oil, and oil is a natural resource that will eventually run out unless new deposits are found. No, the crisis can be avoided if we develop the other sources of energy that are available to us. Of course, developing those other sources takes time and a great deal of money, so oil continues to be our most valuable natural energy source.

Although oil has always been valuable, it has not always been as expensive as it is today. For years the gasoline that was sold at neighborhood service stations cost only pennies a gallon. And then, in 1973, something happened to change the price of a gallon of gasoline forever. To understand the reasons behind the sudden increase in the price of gasoline, you have to understand a bit about the economic background of the oil business. The following is a very simplified explanation.

The oil business, and most other businesses, runs on the economic theory of supply and demand. If the demand for something is much greater than the supply of it, the price of the item goes up. On the other hand, if the item is plentiful and the demand for it isn't too great, then the price goes down. For example, if you were the only person mowing lawns or baby-sitting in your neighborhood, then the supply of your kind of service would be short, and you could perhaps charge more for it. But if other kids in the neighborhood were also in the business of mowing lawns or baby-sitting, then there would be plenty of supply for your kind of service, and the price you charged might have to come down in order to compete. Then, too, you would have to be certain that there was a demand for your business in the first place. A lawn-mowing business won't make you much money if you live in the city, and baby-sitters won't get rich working in a retirement community. In order to compete, there must be a need for the product, and a balance between the supply of it and the demand for it.

The United States is the country that developed the oil business and, in the beginning, created both the supply and the demand for its product. As we have said, oil was at first considered more of a nuisance than anything else. It seeped into the water wells that people were drilling, ruining the water. But when, thanks to Professor Silliman's report, George Bissell saw the commercial possibilities of refining kerosene from crude oil, the oil industry really began.

Kerosene was the product Bissell was looking for, because there was a demand for it. Kerosene lamps were replacing candles as a way to light up homes and businesses. Bissell formed the Pennsylvania Rock Oil Company and hired Colonel Edwin Drake to drill a well. Drake drilled the first commercial

Two kinds of energy are produced side by side.

oil well, in Titusville, Pennsylvania. It was 69 feet deep on August 27, 1859, when the bit pierced the trap and oil began to flow into the well. The oil industry was on its way, producing kerosene from the crude oil. The gasoline produced in the refining process was thrown away—who needed that stuff?

Something happened to change people's minds about both kerosene and gasoline, and once again, it had to do with supply and demand. Thomas Edison invented the electric light bulb, and the demand for kerosene lamps was on its way out. Henry Ford introduced mass production to hold down costs in his automobile factory, and suddenly the possibility of a car for every family was on its way in. Oil was needed to generate the electricity that was lighting up the world, and gasoline was needed to run the cars and trucks and buses that were taking people and products where they needed to go. Oil companies happily drilled more wells to supply the increasing demand for their products.

At first, the United States could supply all of its own demand for the product. Then, as the demand continued to increase, our own supply of oil dwindled. Before long, all the easy-to-find, and easy-to-produce oil in the United States had been discovered, but the demand for the product was still increasing. In 1960, the United States consumed 9.7 *billion* barrels of oil, and 1.6 billion of those barrels had to be imported from other countries. By 1981, the United States was consuming 15.5 billion barrels of oil a year and importing nearly a third of it, or 5.3 billion barrels.

How did we get so dependent on imported oil? There are many reasons. For a long time the United States remained the world's largest producer of oil, and for years the cost of a gallon of gasoline was quoted in pennies. Even though we had to import some of it, we still thought gas and oil were plentiful, and we used them extravagantly. Our cars guzzled gasoline. We burned our lights brighter and longer than necessary. Cars, trains, planes, trucks, whizzed along our highways delivering goods and services to the far-flung corners of the country. We kept our homes warm in winter and cool in summer by using electricity. In fact, we used electricity to power practically everything, from our refrigerators to our toothbrushes and can openers. You could say that, in this country, we had become almost addicted to cheap energy, and that was a mistake. Why? Because the United States could not produce enough oil to meet its own energy needs.

Up until 1938, the United States was not only the world's largest oil-producing country, it was also an oil-exporting country. In other words, it had more oil than it was using, so it was able to sell oil to other countries that didn't have enough or had none at all. In 1938, Venezuela took over as the world's largest exporter of oil, but the United States still produced more oil than any other country in the world. Eventually, however, the oil in the United States that was easy to find began to be used up, and the oil companies looked elsewhere for places to spud in and begin new wells.

The geologists and geophysicists who worked for American and European oil companies knew that oil was located in traps beneath the surface of many different countries. They wanted to drill wells in oil-rich countries like Iran and Iraq, Saudi Arabia, Libya, Venezuela, and many others. Although these countries had rich deposits of oil, they did not know how to go about looking for the oil, drilling for it, or bringing in the wells. They agreed to let the well-established American and European oil companies find and produce the oil in their land in exchange for a flat fee. A flat fee is a one-time payment for something. A *royalty*, on the other hand, is a share of the earnings that something—such as a book, or a play, or an oil well—makes. If the well produces a lot of oil, then the amount of royalty that is paid is substantial. If the oil well, or the book, or the play, does not sell or produce much, then the royalty paid is small, perhaps even nothing. Oil companies that drill in the United States pay royalties to the person or groups of people who own the land where the oil well is located. However, during the time we are talking about, oil companies paid flat fees to the governments of the oil-rich countries in other parts of the world. After the fees were paid, the oil companies were free to sell the crude oil any way they wished. Because of these agreements, by the end of World War II there were giant oil fields in production in the Middle East. By 1976, Saudi Arabia had taken over the number-one position as the world's largest producer of oil. However, the United States was still the world's largest consumer.

The European countries, the United States, and the Soviet Union are referred to as industrialized nations—meaning that they have scientists and businessmen and artists and craftsmen who keep their countries up to date in all the latest technological developments. Countries that are underdeveloped or that are just beginning to develop their natural resources are called developing nations. Sometimes they are referred to as "Third World" countries. Eventually, the developing nations began to feel that the European and American oil companies had not been paying them enough through the years for the right to take crude oil from their land. Their argument was, "It's *our* oil, and you're making most of the money from it!" The oil companies, on the other hand, felt that they were paying a fair price and argued that the oil could not have been produced at all if it weren't for their technical knowledge of exploration and drilling. The oil companies also argued that, since they had paid all the tremendous costs and taken all the risks of drilling the oil wells in the first place, the profits they made were reasonable.

The arguments were not easily resolved. Some of the developing countries decided to band together. They formed an alliance called the Organization of Petroleum Exporting Countries—OPEC for short. Some of the governments in countries that belonged to OPEC decided to take over the oil operations in their land, and they nationalized the oil companies there. That means they simply took all of the assets —the drilling rigs, the housing, the money, everything—of the oil companies. The oil companies had no choice but to leave the country, and all of their employees and their families moved out. Other OPEC nations offered to purchase the assets of the oil companies—however, the OPEC nation, not the oil companies, set the price for the sale. The companies had no choice but to agree to the terms of the sale, and they took their money and left. Still other OPEC nations wanted a majority share (more than half) of the oil companies' operations, but they wanted the oil companies' people to stay and run the business. In many cases, the oil companies agreed to this arrangement.

No matter how the countries gained control of the oil in their land, they agreed they would act as one organization —OPEC—when it came to producing and pricing that oil. Being dependent on others for something you desperately need can be a risky business, as the world found out in 1973. Two-thirds of the world's deposits of oil lie in OPEC countries, so OPEC had—and still has—a powerful weapon that could be used to influence the politics or affect the economy of other nations. In 1973 the OPEC nations met and agreed to *quadruple* the price of a barrel of oil and, at the same time,

Oil wells sprout in the deserts of Saudi Arabia . . .

limit the amount of oil that would be pumped from their wells. Suddenly, oil was scarce. Its price skyrocketed. The economic theory of supply and demand was at work as usual.

There was a general world-wide panic as countries scrambled to get what they needed of the precious fuel. Because the United States could produce two-thirds of the oil it needed, it was not in as much trouble as countries that had no oil reserves of their own and had to import every single drop. And

yet, because the United States used—and continues to use—
more oil per person than any other country in the world, the
oil embargo of 1973 hit hard. Lines formed at gas stations,
and people waited hours and hours to fill the tanks of their
cars. Laws were enacted to lower the speed limit on interstate

. . . while the drilling goes on in America.

highways from seventy miles per hour to the more gas-saving fifty-five miles per hour. President Nixon requested that there be no decorative lighting on homes and businesses during the Christmas season, in order to save energy. Homes were cooler in the winter and hotter in the summer as people adjusted their thermostats, trying to save money on rising heating and cooling costs. Individuals were not the only ones affected by the sudden shortage of cheap oil. Companies that could not afford the higher cost of fuel had to close, putting people out of work. When people have no income, they cannot purchase goods or services, so companies that provide goods, such as car dealerships and furniture stores, or services, such as restaurants or dry cleaners, begin to lose money because they have fewer customers. Fewer customers, fewer profits, and still more people lose their jobs. Times like this can cause the economy of any country to practically come to a stop—a situation that is referred to as a recession.

When the demand is greater than the supply, the world experiences an oil shortage such as the one in 1973–74, and the price of a gallon of gas at your neighborhood gas station skyrockets. The price of a gallon of gas stayed high for several years, and people began to adjust to the new, higher cost of energy. Then, in 1981, the OPEC nations met together and decided to *lower* the price of a barrel of crude oil and, at the same time, provide more of it. People spoke of an oil glut, and the price of a gallon of gas at the gas station began to tumble, but not to the pennies-per-gallon prices of a few years earlier. Those days of very cheap gasoline are gone forever. Because OPEC kept the price of their oil so low, American and Euro-

The search for new sources of energy continues.

pean oil companies and those nations that didn't belong to OPEC could not afford to compete. You have seen how expensive it is to drill an oil well and what the chances are of finding oil when the well is completed. Because of OPEC's price controls, any oil the companies found would cost more to produce and refine than it could be sold for, so profits for the American and European oil companies fell, people lost their jobs, and the cycle repeated itself.

Although OPEC countries control most of the oil on this planet, other industrialized countries will continue to explore for oil in their own lands. Most of this exploration will be difficult and expensive, because the oil that is yet to be discovered is in remote or dangerous locations. Nevertheless, the searching and drilling for oil will continue all over the world.

Meanwhile, scientists and engineers will continue to look for other sources of energy as well. For instance, we know that layers of rock called oil shale can yield a substance called shale oil, and shale oil can be used to fill energy needs. Oil can be extracted from tar sands, too, and scientists are working on a way to turn alcohol into a substance called gasohol that will power cars.

Coal and wood can provide energy. So can the sun, the wind, and water. Nuclear energy and its uses are a subject that is hotly debated everywhere, and scientists are looking into the possibility of harnessing geothermal energy, too. (Geothermal energy is the kind of energy that makes geysers like Old Faithful perform.) In Brazil, plantations of the Cassava plant, or oil tree, are being grown. Its bark produces a kind of synthetic oil. There is an energy crisis in our world, but it is not a crisis we cannot solve. Because of short supplies and great demand, the people of the world must work together to find the energy we need today—and will need even more of tomorrow.

Index

Grateful acknowledgment is made to the following for their contribution of photographs:
    Brown Brothers: photographs on pages 24–5 and 48.
    Exxon Corporation: photographs on pages 4, 11, 21, 23, 26, 27, 30, 32, 36, 38, 42, 50, 51, 53, 55, 56, 57, 58, 60, 63, 64, 67, 70–1, and 82.
    International Association of Drilling Contractors: photographs on pages 9, 12, 13, 14, 15, 16, 17, 19, 34, 35, 40, 41, 44, 77, 83, and 84.
    Western Geophysical Company of America: photographs on pages 6 and 7.